Traffic and the environment

a paper for discussion

Contents

1 Introduction

1.1   The improvement of London's system of communications, whether by road or by rail, involves complex and lengthy procedures and long construction periods. While considerable progress will be made in the next ten years, the effects will benefit particular areas rather than the whole of London. Meanwhile the pressure of growing car ownership and the effect on the environment will increase as it has been doing over the last twenty years.

1.2   This pressure has been controlled to some extent by such measures as parking controls and traffic restriction. If, as seems likely, the pressure increases should the existing methods of control be strengthened or should other methods be tried? This paper discusses some of the possibilities. It is not a statement of Council policy, but public comment on it will help the Council to decide what it should do, and what it should urge other public authorities to do, on a matter of vital importance to London's well-being, the balance between traffic and the environment.

2   The general shape of the problem

2.1   We have all lived with the problem of traffic and its effect on our living and working conditions for a long time. Much has been said and written on the subject; much more will no doubt follow. In spite of this familiarity and perhaps because of it the steady growth in the number of motor vehicles and their use and the effects of both on London's life are not as keenly noticed and understood as they might be.

2.2   The number of cars in London is still growing despite there being less people and fewer jobs; the volume of traffic has grown steadily at the rate of about five per cent a year for the last twenty years. Over most of London it shows at the moment every sign of continuing at this rate. Figure 1 shows the trends.

2.3   Since there has not been much new road building in London in the last twenty years our inherited haphazard network of roads has had to

meet the strain. It has done so surprisingly well. Improved vehicle
performance and traffic schemes have, if anything, led to higher speeds

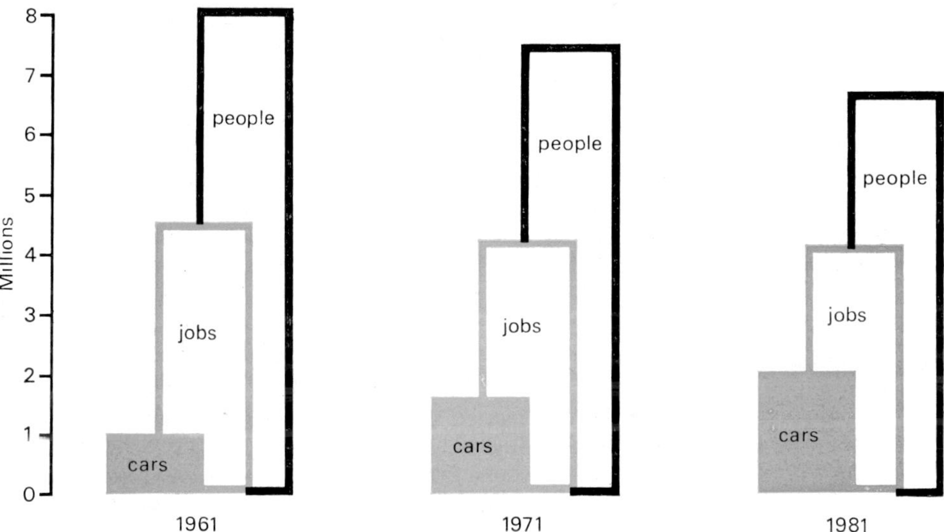

Fig. 1   The changing balance of people, jobs and cars

and greater safety for vehicles on the roads than ever before. This surpris-
ing situation may well have lulled many Londoners into a state of false
security—a feeling that this process can go on forever.

2.4   But of course it cannot. The margin of spare capacity in the whole
system is being eroded continually and there must come a moment when
the limit is reached. In central London we have now come very close to
the limit and in the next ten years much wider areas could reach a critical
state in which the whole level of convenience and service given by the
roads declines sharply, with a damaging effect on the pattern of the daily
life, work and recreation of many Londoners.

2.5   Even more threatening are the consequences of such continued
growth to the quality of life in this coming decade, during the time when
road improvements will not have had much effect. Increasing congestion
means busier conditions on more roads, more traffic in the evening and at
night when people most want peace and quiet and more traffic on local
streets. It means less reliable and slower bus services. It means more
noise and more pollution.

2.6   This subjects pedestrians to great inconvenience and danger. Difficulties in crossing the high street can spoil many local shopping centres; getting to school can be unpleasant and hazardous. About eighteen thousand pedestrians are injured and four hundred killed each year on London's roads. Almost two-thirds are children and old people. One-third of the accidents occur on local roads.

2.7   The Council saw these problems when it put forward its Development Plan. The strategy it propounded was that general development in London, whether of offices or shops, houses or parks, large areas or single buildings, should be planned in relation to the present and developing transport system. Within this general principle the Council plans to improve public transport, build more roads and to restrain the use of cars. How effective can this long term strategy be in the next ten years?

2.8   Any improvements in the relationship between transport and development in general will be long term in their effects. Even general measures not specific to the particular problems of London such as the better insulation of buildings against noise or the design of quieter motor vehicles would not of themselves be sufficient. This is not to say that the latter measure especially should not be pursued with great vigour by Government and manufacturers as an approach offering good return in environmental improvement for money spent.

2.9   Improvements in the quality and reliability of public transport can help to relieve the situation but there is no likelihood that they would reduce car use sufficiently to make a really substantial impact on the growing problem. The use of cars has grown only moderately by the transfer of journeys from bus or train but overwhelmingly by extra journeys encouraged by the greater convenience and flexibility of the car. Nevertheless the standard and availability of public transport services is an important factor in making other, more direct, measures of car restraint acceptable. The particular problems of buses in the traffic/environment equation need special attention.

2.10   Major road construction will in the long term have a considerable impact on the problems and indeed many will never be satisfactorily dealt with without such action. Some places are already benefiting from new roads and more will as the programme gathers momentum. But in the next ten years what can be done will not give sufficient relief from traffic congestion to provide opportunities for environmental improvement over the whole of London.

2.11   It is impossible to escape the conclusion that in the short term the problems of traffic and the environment can only be attacked effectively by direct measures to restrain the use of motor cars. The prospects raised by increased restraint are, however, neither easy nor straightforward. Cars have brought not only problems but immense benefits; commercial vehicles efficiently used have reduced the cost of goods and services.

2.12   To what extent should improvement of the environment be at the price of reducing these advantages? The livelihood of London depends on the road system and the motor vehicle. Eighty per cent of passenger journeys and over ninety-five per cent of goods traffic go by road. Over half the families in London own a car and expect to use it; they might be tempted to join the ever-increasing numbers leaving London if they are not allowed reasonable use of it.

2.13   The main road system provides the circulation necessary to keep London going and any substantial restriction of that circulation would have important consequences.

2.14   Up to now many improvements to traffic and environment have been possible without much hardship to anyone. In future environmental gains to some will involve restrictions for others. What some want others will not. The Council will in the end have to judge what it is best to do. The choices will not be easy but they must be identified and discussed.

2.15   In the following chapters the main opportunities for action are presented. For each the assessment must be made—how effective is it likely to be in dealing with the situation we face in the short term and how acceptable is it?

3   Parking

3.1   Londoners are now familiar with control of parking on the street. Originally such controls were introduced primarily with the object of tidying up the street system, of clearing away parked cars where they got in the way of moving traffic and of making the streets safer. But it did also start the process of rationing the road space for the various types of user. This has been successful so far as it has gone.

3.2   But the Council's parking policies also seek to restrain traffic in congested areas by control over all forms of parking: it is the total amount available both on-street and off-street which matters if restraint is to bite. How effective can this approach be in this decade?

3.3   In spite of the considerable efforts made by the Council and the London boroughs in introducing on-street control there still remains much that might usefully be done. Control in inner London needs extending; more suburban town centres could with advantage be controlled. Table 1 gives some idea of what could be achieved in the Inner London Parking Area—the forty square miles of greatest congestion. Outside it fourteen town centres are already controlled and this number could at least be doubled in the next ten years.

Table 1—On-street parking control in the Inner London Parking Area

	1961	1971	1981
CENTRAL AREA			
Number of meter spaces	10,500	21,000	20,000
Number of residents' spaces	—	7,000	9,000
Number of free spaces	53,000	2,000	—
TOTAL	63,500	30,000	29,000
REMAINDER OF ILPA			
Number of meter spaces	—	12,000	30,000
Number of residents' spaces	—	26,000	60,000
Number of free spaces	210,000	134,000	30,000
TOTAL	210,000	172,000	120,000

3.4   What about off-street parking? Clearly when there is a problem of fitting a growing traffic demand to a limited road space a very wary eye should be kept on the provision of new car parks. Too many would encourage traffic and make road conditions worse. Indeed the money they cost to build would be ill-spent if the cars could just not reach them in sufficient numbers to fill them.

3.5   The Council has already adopted policies designed to limit the growth in the number of off-street parking spaces. In particular it said firmly in 1969 that any more private spaces provided during the construction of new offices should be subject to a strict maximum. Ranges have been prescribed within which the London boroughs should control the situation. This important decision has been written into the Development Plan. The present outlook suggests strongly that the boroughs would be advised to allow only the minimum provision in most cases.

3.6   In much the same way it seems right that the quantity of public off-street parking should be limited, particularly where the existing stock of available parking is high and congestion appreciable, such as in the Central Area and the busier suburban centres like Croydon. But surely it would be wise of all authorities to take a realistic view of their situation and not provide for parking, and therefore for traffic, beyond what their neighbourhoods can clearly accommodate now.

3.7   There are other possibilities for controlling parking besides those relating to the quantity which can be allowed, particularly where there is a great need to limit the traffic. The Council has said in the Greater London Development Plan how it thinks methods of traffic restraint should be judged. They should:

 i) control journeys causing most traffic and environmental nuisance

 ii) limit those car journeys which easily transfer to public transport

iii) limit the less important journeys

iv) restrict as few journeys as possible whilst achieving the desired effect

 v) be thought fair and reasonable

vi) be administered and enforced without too much difficulty.

3.8   The present parking control policies unfortunately do not score highly against these criteria. Public car parks operated in such a way as to encourage peak-hour travel fail to satisfy the first two criteria. Lack of control over the use of private parking under offices and the general inability to deal with through traffic mean that none of the first five criteria is satisfied. The difficulties of enforcing on-street parking control lead to the sixth not being met.

3.9   The problem of enforcement is crucial to any substantial extension of on-street parking control as well as to any other traffic regulations which are not self-enforcing. Indeed the Council first went on record as long ago as 1966 to say that changes in the law were required which should include *keeper liability* for minor traffic offences, and it still holds strongly to this view. Each vehicle would have a named keeper (usually the legal owner) who would be responsible, subject to reasonable safeguards, for the payment of fines for those traffic offences now dealt with by the issue of tickets. At the moment it is the person who was driving at the time the offence was committed who is responsible. A fairly large number of people ignore the rules at present because they find that even when they are given a ticket the cumbersome nature of the system allows them to get away without penalty.

3.10   With the idea of being able to restrain traffic more effectively the Council took powers in section 36 of the Transport (London) Act 1969 which would allow it, with the boroughs, to designate areas within which public off-street car parks would need a licence to operate. The licence could contain conditions affecting the use of a car park, such as the scale of charges, opening and closing times, and the duration of parking allowed. A study of part of the Central Area has now been made and the implications of a licensing scheme there explored.

3.11   The study area is shown in Figure 2: it contains almost half the Central Area's shopping and about a quarter of its office space, as well as about fifty per cent of its public off-street parking. About two-fifths of the public off-street parking space in this area is let on a contract basis and over eighty per cent is charged for on a daily commuter basis. At present these spaces generate about seven thousand car trips in the morning peak (0800–1000 hours).

3.12   Licensing would not reduce the total supply of public off-street parking in the area but would change the charging structure. Many car parks currently charge on a flat-rate, all-day basis. To change this system of charging to one based on an hourly rate would mean that the all-day parker would incur heavier charges than now.

3.13   Such controls could almost halve the number of commuters using these car parks; the number of cars parking off-peak would increase. For the daily commuter who values the use of his car highly, parking would still be possible but more costly. For the shopper, businessman and visitor travelling outside the peak, parking would be easier and possibly not much more expensive than at present.

10

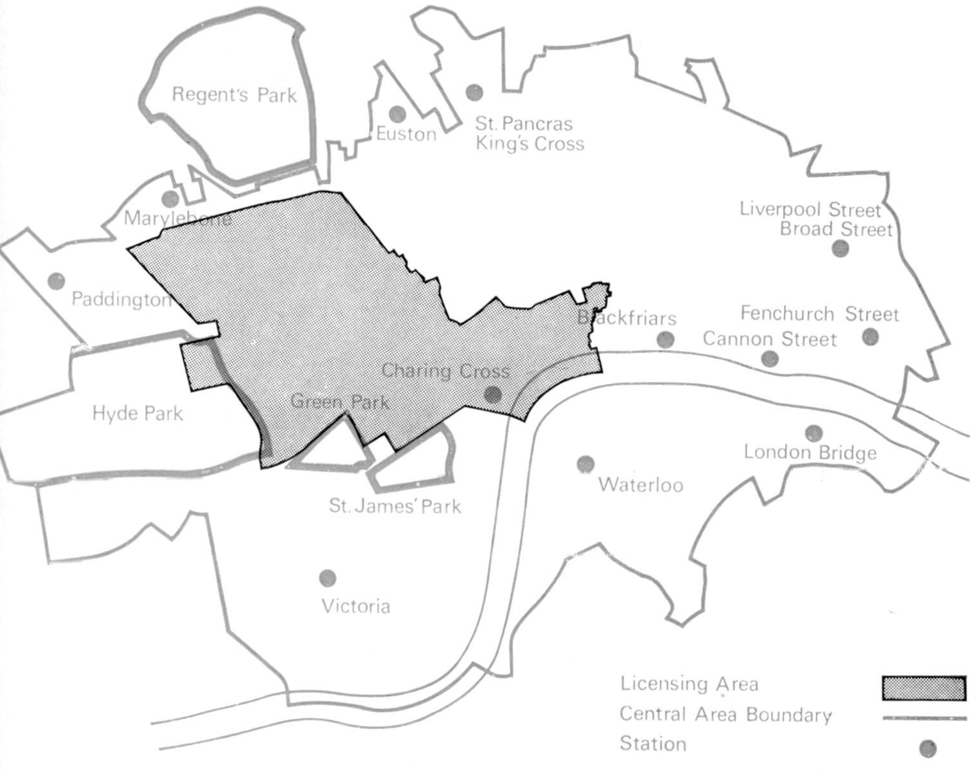

Fig. 2 Proposed licensing area for public off-street car parks

3.14   Even if all public off-street car parks could be controlled in this way to reduce congestion in the peak hours and its effect on the environment, increased use of the large pool of private parking space, particularly in office buildings, might offset the gains. Table 2 shows that more than half (53,000 spaces) the off-street parking in the Central Area is in such buildings. Clearly this parking is an important factor in the situation.

3.15   That this space exists in such quantity is very largely due to earlier planning requirements which made minimum provisions for car parking mandatory on developers of new buildings. This was thought to lessen the parking load on the streets and thereby ease traffic conditions. What was once a strict requirement and involved building owners in extra cost and complexity (but has none the less been found useful by the occupants) is now seen to have been misguided. The Council's new policies are therefore to limit stringently any further increase. What should be the attitude towards the existing spaces? How many spaces should be provided for regular journeys to work? How many will be needed for operational reasons, particularly to accommodate callers during the day? It is the excess over these operational needs to which attention should be directed.

Table 2—Parking supply in Greater London 1971

| Area | Off-street | | | | On-street | Total |
| | Public | Private | | Total | | |
		Residential	Non-residential			
Central	30,000	17,000	53,000	100,000	30,000	130,000
Inner	20,000	180,000	120,000	320,000	680,000	1,000,000
Outer	70,000	600,000	350,000	1,020,000	1,400,000	2,420,000
Total	120,000	797,000	523,000	1,440,000	2,110,000	3,550,000

3.16 Clearly it would be preferable if as many spaces as possible which are surplus to operational needs and now largely used by commuters could either be operated as licensed public car parks or converted to some use other than parking.

3.17 These private garages vary in size and many would be too small to make suitable public car parks. In the Central Area for instance three out of every four private garages in office blocks have less than fifty spaces each. Private garages as small as this could perhaps best be converted to other uses, at least in part.

3.18 But what scale of change might be desirable? If over the whole of London three thousand parking spaces in private buildings could be converted each year, one-third into public car parks and two-thirds into, say, storage space, this would create another half-million additional square feet a year in commercial use. But this need not distort the Council's office floorspace policies if administered carefully.

3.19   Some such changes could be achieved by agreement. If the size and layout of a building were suitable owners might be persuaded to convert their private garages to public use or to storage. Councils could acquire them by agreement and run them as public car parks. But now the habit of using these parks for commuting has become established many owners might be unwilling to change voluntarily. Is the situation sufficiently serious for the Council to seek powers to purchase such space compulsorily?

3.20   This might not be either a welcome or equitable operation. If stern measures were thought to be necessary might it not be better instead to institute a tax on all parking in this category? To be effective the levy might have to be high—anything from £50 to £250 per space per year in central London. A tax of this kind might encourage the present occupants to give up their private spaces in favour of public use or convert them to storage. It might also be regarded as an incentive for developers of new buildings to provide less parking than the maximum permitted.

3.21   If we are to try seriously to alter the existing traffic and environment balance it seems inescapable that present patterns of use of private non-residential parking space in central London, and perhaps in the largest suburban centres too, should be changed. The idea of compulsion may well be repugnant to many. But unless there can be very substantial voluntary moves the possibility must remain. The Council for its part is reviewing the use of its own car parks. It would be a considerable help if Government departments and other large organisations in the Central Area would do the same.

3.22   Any such new measures must take into account the value of different parts of London as places to live in. Controls which improve the traffic and environment in an area will be of little value if the residents who should benefit move out because the controls are too restrictive and the area becomes simply a pleasant and convenient place to drive through or visit. Provided the traffic they generate is not too great at busy times residents should have a fair measure of priority—certainly in preference to commuters.

3.23   What is the future for parking controls? With little immediate increase in road capacity in prospect restraint will need to be tougher and the limits which parking can put on car use are likely to be essential. How far the measures will need to be taken is a matter for consideration. The extent must depend on how badly the community wishes to change the status quo and on whether there are other restraint measures which would be preferable or would limit the need for a very much tougher parking control.

3.24   The risk is that the more complicated and controversial alternatives might not come into force quickly enough. There is above all the problem that parking controls in an area deal only with traffic which ends its journey there and not with through traffic. In the Central Area for example one vehicle in five merely passes through and in some surburban centres this proportion is as high as two in three. Other methods should, therefore, be considered as possible complements or alternatives to further development in parking control. These are considered in the next chapter.

4.   New methods of traffic restraint

Supplementary licensing

4.1   A supplementary licensing scheme would require the purchase of a special licence or ticket, in addition to the annual road licence, in order to use a vehicle at specified times in a designated area. This method was considered by a Working Party set up by the former Ministry of Transport in 1965 and was discussed in their report *Better Use of Town Roads* published in 1967. They concluded:

> " We are far from satisfied that it would be practicable to introduce some system of supplementary licences or admission charges for using vehicles in congested areas, and we are, in any event, doubtful whether in practice a system of this kind would necessarily have advantages over comprehensive parking control. The feasibility and practical effectiveness of these systems as compared with parking controls could not be assessed without experiments requiring legislation." (*Better Use of Town Roads*, Para. 1.2.(6))

4.2   Five years have passed since this conclusion was reached. During this time the difficulties of parking control have become more clearly understood and the need for effective action has grown. Furthermore the continued development of parking policy seems itself likely to be complex and difficult. Would it not then be right for the Council to consider

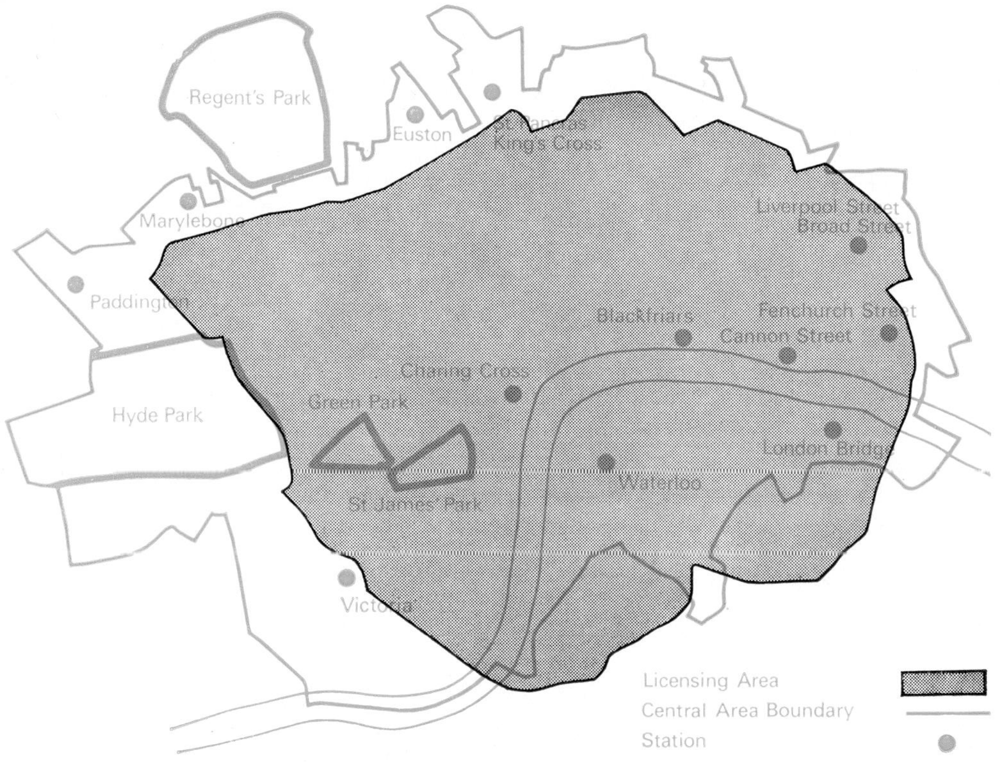

Fig. 3   Possible supplementary licensing area

supplementary licensing afresh and to work out in some detail what might be necessary?

4.3   One can see at once that the most obvious candidate for the initial introduction of a supplementary licensing scheme is, broadly, central London itself. A possible area of control is indicated in Figure 3. If a licence were required to use a vehicle in the area between 0700 hours and 1800 hours on Mondays to Fridays this would not interfere with evening and weekend leisure pursuits. It could however improve the daytime environment and reduce congestion during business hours. Those who did pay would benefit by the freer movement of traffic. The improved service in public transport to central London would be a good cushion for car drivers who elected not to pay.

4.4   What might the details of the scheme be? Tickets would have to be displayed next to the ordinary road licence and there would have to be systematic spot checks on the roads in the Central Area. Tickets could be supplied from post offices and vending machines or perhaps from special licence offices. Some people would certainly wish to have season tickets, others to have the opportunity to purchase a daily ticket. Penalties geared to the cost of a daily licence would be needed to discourage evasion.

4.5   It is difficult to estimate just how much the charges might be. Experiment would be necessary to find the level which produced the right degree of restraint. It would be feasible to aim at traffic reductions of up to a quarter at peak periods to gain an appreciable increase in traffic speeds. Any increase in general traffic speeds is important for buses. There would be some scope for dealing differently with different vehicles. Public service vehicles could be exempt and a difference might be made between goods vehicles and cars to reflect the different environmental effects of each. Local residents would need special consideration. There is no overriding reason why they should be excused the fee completely but some concession might be desirable. After all, London is a place to live in as well as a business centre.

4.6   Supplementary licensing in the Central Area would have important effects immediately outside it. On the one hand it would divert a proportion of today's through traffic from central streets to other streets in inner London; on the other it would reduce vehicle movements on the way to and from the Central Area. The net effect might still be an increase in traffic in inner London, particularly around the edges of the Central Area. Two complementary measures would be needed to deal with this problem. The first would be a buffer zone around the control area to prevent over-parking around the boundary; the second would be the definition of by-pass routes so as to provide alternatives for through traffic.

4.7   This points to one of the problems of supplementary licensing in the absence of good by-passes in the area involved—the shift of the weight of traffic, and therefore the environmental repercussions, from one area to another. In the short term these difficulties would have to be weighed carefully; in the long term a road like Ringway 1 could avoid them. The other clear problem of a system like this is the administrative complexity of issuing the licences and enforcing the law. Once again, whether these would be acceptable must depend on how much everyone would value the result.

16

4.8   If a workable arrangement were to be established in central London, it is possible to imagine the system extended so that the more congested areas of London would be covered by peak-hour licensing only, with central London and perhaps one or two centres having additional all-day sanctions. But this complexity invites one to ask whether there might not be in the longer term a more flexible and sophisticated system if indeed the need for further measures is proved.

Road pricing

4.9   One such possibility has been mooted already—direct road pricing. In principle it is simple. Every vehicle would be so equipped that its use could be metered, the unit charge being varied to reflect the degree of congestion on the road system so that journeys at busy times and in busy areas would cost more than off-peak journeys on quiet roads. In practice it might well be expensive to install—there are two million vehicles in London alone to deal with—and it might not reproduce readily on the ground the fineness of control that some of its proponents have in mind.

4.10   Some five years ago *Better Use of Town Roads* concluded:

> " Direct road pricing of the kind described in the Smeed Report is potentially the most efficient means of restraint. It would take several years to develop and put into effect. Although it is by no means certain that a workable system could be devised its advantages would be so great as to justify further research and development work on it."
> (*Better Use of Town Roads*, Para. 1.2(9).)

4.11   Since then some development work has been done by the Government following this recommendation. There has been little recent comment in public on the outlook for road pricing. It is not difficult to imagine just what complexities of electronics and administration might be involved nor to see how controversial it might be. It is very much a development on the national plane and for the Government to consider.

4.12   There are many problems yet to be solved and road pricing may well be a possibility for the next decade rather than this. But if extensive controls were necessary road pricing could offer a means of restraint which is more efficient, flexible and fairer than parking control or supplementary licensing. The problems of other means of restraint suggest that the further development and evaluation of road pricing should continue so that the long term options are kept open to deal with these extremely difficult and only partly-understood matters. If this were thought right it would be for the Government to take the lead, the Council giving help with its special experience of complex traffic problems.

5.   The general management of roads

Main roads

5.1   For this decade, in many parts of London the existing main roads must continue to provide for cross-London journeys, for local circulation and for many of the more important bus services. The better they function the more they protect the quieter roads in residential areas from intrusive traffic. The Council sees the principal management aims of these roads as being to:

 i) increase traffic safety

 ii) reduce pedestrian accidents

iii) reduce delays and congestion

 iv) facilitate the exclusion of extraneous traffic from local roads

 v) provide priorities for buses and, in some circumstances, for taxis.

5.2   Many of the traffic management measures are familiar—one way streets, banned turns, entry prohibitions, urban clearways and the like. The GLC and the London borough councils, with the co-operation of the Police, have used all these techniques widely in the last few years and have achieved a considerable improvement in the safety and smoothness of traffic flow. The difficulties of enforcement under present law however limit the effectiveness of many of them. Accident rates have been reduced in spite of the growth of traffic. For the future the potential of the more familiar measures is limited but in other directions much remains that can usefully be done. Bus priorities are discussed later. Computer traffic control, selective road closures and improving the scene along main roads are touched on here.

5.3   The Council is well advanced with computer traffic control. The linking of 1,070 traffic signals in London with a central computer to get the best possible setting of the signals is the target. By the end of this year traffic signals will have been added to the 70 covered by the experimental scheme in west London so as to complete control of the busiest area. Computer traffic control will give a once-and-for-all improvement in the performance of the road system perhaps equivalent to an effective increase in capacity of about ten per cent—not to be sneezed at when physical improvement can be so difficult and damaging. There will be opportunities for experiment with signal settings to favour buses and to control the entry of traffic into selected areas. The more widespread provision of pelican pedestrian crossings will become easier.

5.4   The selective closure of side roads is worth considering. Frequent connections between local and main roads can be environmentally harmful because of the opportunities they give for traffic to use back streets to by-pass congestion on the main roads. Turning traffic also creates hazards and slows the traffic on the main road. An analysis of the A10 and A1010 between the North Circular Road and the Greater London boundary shows how accident rates are linked closely to the number of minor junctions on a main road. Table 3 illustrates this.

Table 3—Accident rates and junction frequency (1970)

	Accident rate (injury accidents per million vehicle miles)	Frequency of minor junctions (junctions per kerb mile)
A10	2.3	3.5
A1010	5.0	9

5.5   These figures, which are fairly typical, suggest strongly that the selective closure of minor junctions on main roads would be of substantial advantage. There would be gains for traffic also: a comparison of these two roads shows that speeds and flows on the A10 are almost fifty per cent higher than on the A1010, whilst the carriageway is only twenty-five per cent wider.

5.6   Latest methods now help the Police, the GLC and the London boroughs to gauge the extent of the problem accurately. The computer can superimpose on a map the locations of all the accidents of a specified kind in the way shown in Figures 4 and 5. Detailed study of such maps as these can pinpoint the roads where closely spaced junctions result in a continuous chain of accidents. They can also show where these junctions are related to abnormally high pedestrian accidents in the side streets.

5.7 There are clearly opportunities here but how far the process of closing junctions should go will need careful judgement. It will be a mixed blessing. Many people will find access to the main roads less convenient and delivery vehicles could have difficulties. The reduction of traffic on some side roads might be matched by an increase on others. Concentration of turning vehicles at a few places might lead to added congestion on the main roads and a renewed increase in pressure on local roads. So the argument again is one of balance.

5.8 Conflicts of interest between one area and another are also likely to arise around many town centres if efforts to divert traffic from the main roads running through them are contemplated. If they had to choose, most people would probably think that quiet areas in which to live are more important than improved surroundings for shopping. There may therefore be little choice in traffic terms for most main roads over the next few years. But they must not be written off in environmental terms. A great deal of activity centres on them. More attention must be paid to the needs of the pedestrian, to control of the clutter of road signs and to the opportunities for minor changes in landscaping. High standards of maintenance of the actual road surface is vital but more attention must be given to verges, trees, shrubs, footways and street furniture if there are to be improvements in the urban scene.

5.9 Traffic-free areas for pedestrians, particularly in shopping centres, are deservedly popular. But a close look at the places in Britain and Europe where the best results have been achieved shows clearly that success depends not so much on stopping traffic but on accommodating it satisfactorily elsewhere. Purpose-built precincts are common and often where an existing centre is made free of traffic there is nearby a ring road or by-pass. Progress in London in the short term has been and will be inhibited by lack of such facilities. The shopping streets which are relatively quiet and unimportant for traffic can and should be tackled—Carnaby Street for instance has been freed of traffic. Even where complete closure is too difficult there may be opportunities for severe limitation of traffic for part of the day or even part of the week, say on Saturdays; but experience shows that although this may be a step forward, half-way schemes of this kind have only a limited success.

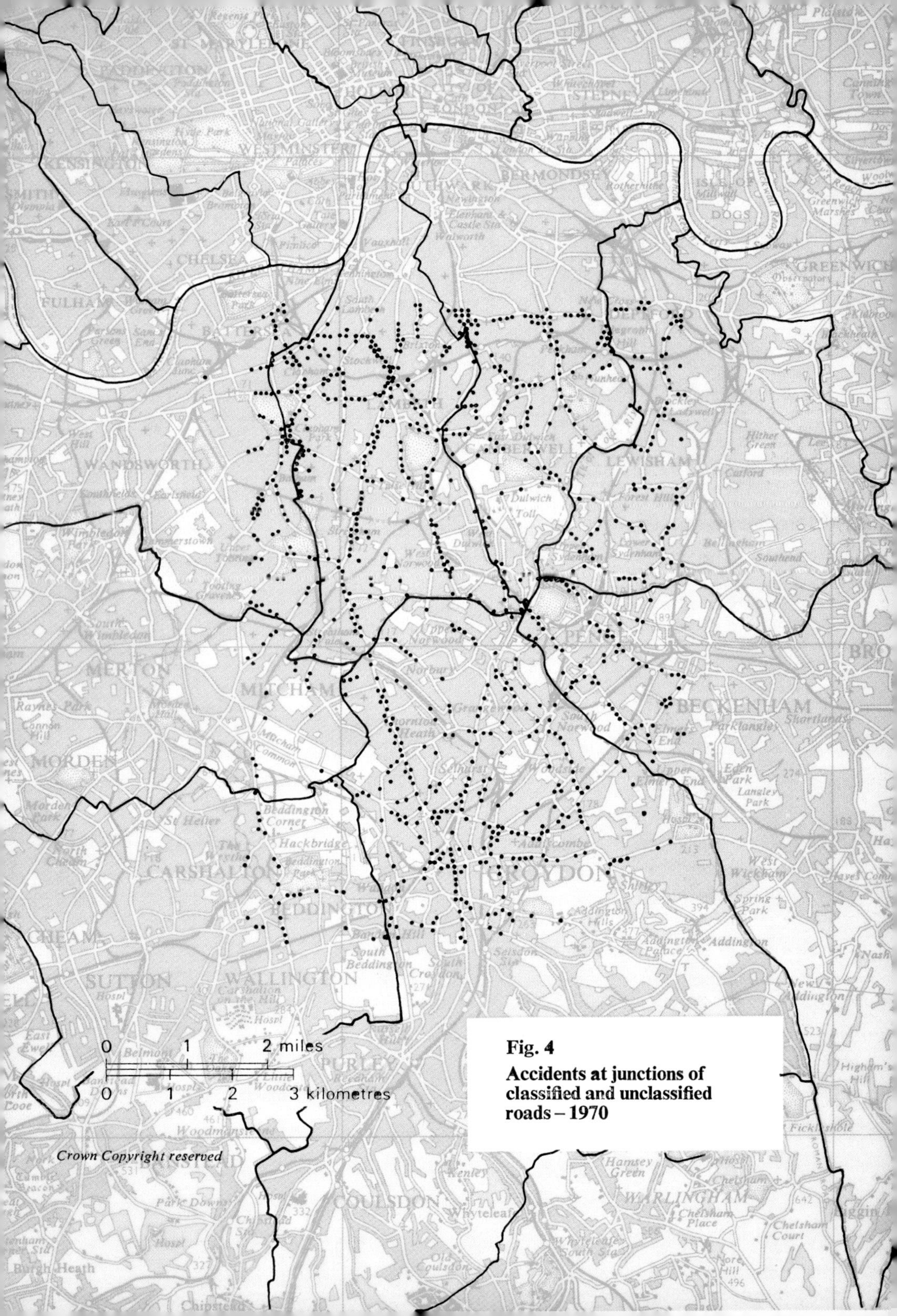

Fig. 4
Accidents at junctions of classified and unclassified roads – 1970
0 1 2 miles
0 1 2 3 kilometres
Crown Copyright reserved

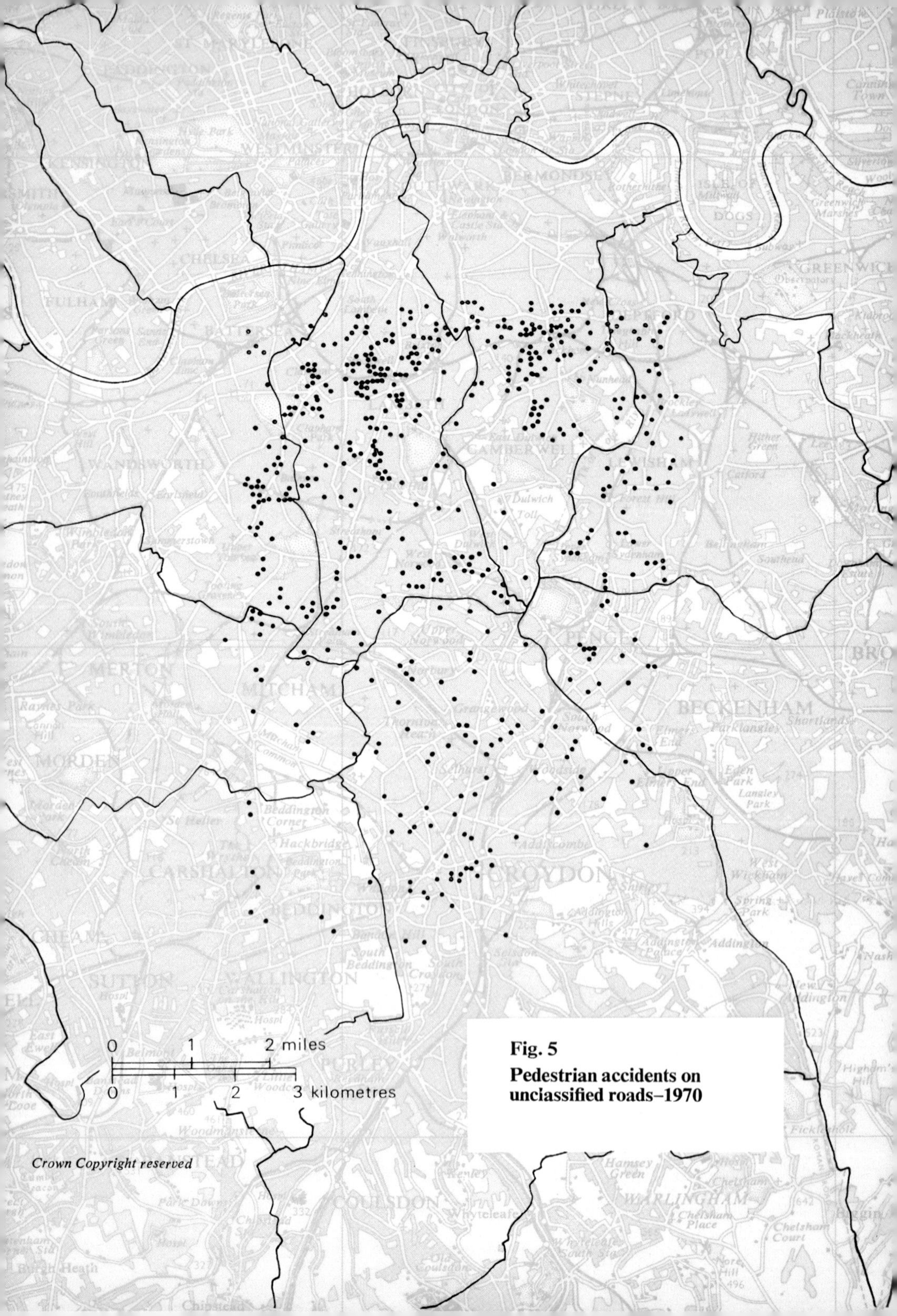

Fig. 5
Pedestrian accidents on
unclassified roads–1970

Crown Copyright reserved

5.10   Many of the varied traffic measures discussed in this paper could lead to improving the environment of the local roads—the roads on which most Londoners live. But in practice, to protect and improve the conditions in residential areas will require careful local study so as to mobilise not only traffic changes where practicable but also the improved maintenance and better layout which are part of the comprehensive approach. Indeed many of the things which give character to local areas are inherited, such as buildings or trees, or created by local residents, such as gardens. Only those who live in an area can put the essential touches to it. Much of the initiative will rightly be local but it is important for the strategic authority to give a lead and do what it can.

5.11   Not everyone will agree on what should be done even as regards traffic. Experience in London to date includes places such as Pimlico where the initiative by the City of Westminster has been successful and accepted. In Barnsbury what has been a help to some appears a disadvantage to others. Possibly the shortage of road space is so acute and the layout of the roads we have so awkward that the room available for manoeuvre is too small to cater for everyone. It is important to keep trying, for only by effective action at local level can the full benefit of wider measures such as new roads be realised. The studies of Greenwich and Blackheath carried out for the Council by Professor Buchanan show clearly the sort of approach which is needed. In the end the precise nature of what is done will vary from place to place according to local circumstances and depending on the balance between accessibility and the peacefulness desired by the residents.

6   Buses and lorries

Buses

6.1   Current opinions about the treatment of buses on the roads range from one point of view which would give them virtually absolute freedom and priority everywhere to another which would consign them to take their chance with all other road traffic. Where in the spectrum of possibilities should the balance be set?

6.2   One should ask first what could be gained by improving the reliability and convenience of bus services. Again it must be said that in relation to the particular central problem which this paper seeks to discuss, the direct effect of improved services on transfer of passengers from motor cars would soon be lost in the current vigorous growth of car use. The Council has therefore emphasised help to buses in order to benefit the many Londoners who use these services regularly and to encourage them to continue doing so, rather than in the expectation that there would be a large scale switch from cars. What is done in this direction must be set against any other effects on Londoners. The more substantial the allocation of road space to bus services the less room for manoeuvre will there be in dealing with other traffic and the associated environmental problems.

6.3   When the traffic on a main road is flowing freely buses, being rather slower than the majority of vehicles and having to stop more frequently, are not impeded and can make reasonably good time. This emphasises strongly that the best help to buses would be a form of general restraint which keeps traffic well within load capacity. But in practice there are also local problems and it is worth reviewing the measures which are in use or available to deal with them.

6.4   Bus lanes enable buses to be freed of delays affecting general traffic. Nearly all the problems come at junctions and the benefit to be gained by long lengths of road set aside for buses away from junctions is not great. The Council is tackling the matter systematically by concentrating on the busiest and most heavily congested bus routes and giving priority to buses where it does the most good. Over thirty bus lanes will be operating by the end of the year and this number will be more than doubled next year. All these schemes will benefit places where relatively frequent buses carry substantial numbers of people. They are unlikely to cause really serious consequential congestion.

6.5   The priority intended for buses in Oxford Street is the most striking proposal so far. As a street it is relatively narrow and includes a whole string of junctions. As a bus route it is one where the buses themselves are a substantial proportion of the total through traffic. As a focus for travellers it is outstanding. Put these considerations together and it is well worth trying out the radical measures proposed. It might well save as much as a million bus passenger hours a year.

6.6   Few schemes are likely to produce such spectacular results and, while this phase of development will not exhaust the possibilities, the present picking out of the plums will mean that the gains from subsequent schemes will be progressively less. What should the guide-lines be? Place and time of day? Buses coming into central London at the morning peak are heavily used and need help. Number of buses and number of people? It seems no more than common sense that only where buses are reasonably frequent and well used would people think that substantial consequential restrictions on other traffic should be contemplated. This is important, for the next phase of development must clearly include a move towards a basic pattern of bus schemes, over much of inner London and in some suburban centres, and progressively there is bound then to be some noticeable interaction with general traffic flow. Indeed the Council is studying the likely effects of using bus lanes in a systematic way, as deliberate chokes to traffic flow in a ring around central London. The disadvantages of added congestion will have to be weighed carefully against possible gains to buses and to restraint in the centre.

6.7   There is a lot at stake here. The desire to help buses, achieve traffic restraint and improve the environment must be tempered by a sober realisation of the importance of the present general traffic circulation. It is not easy to sum up accurately and in advance all the effects that might be produced. The Council's approach has been to tackle the problems as a whole over a wide area with definite priority where action is needed most. What is in hand is far reaching but is being progressed step by step at a pace that allows people to adjust to change and particular problems to be dealt with as they arise.

6.8   There are other on-street measures which are being followed up to help buses. Parking at many bus stops will be banned and more meter schemes and clearways will also help. As the control of traffic signals by computer spreads a bias in favour of bus routes might be built in.

6.9   Although this discussion of buses and their running can be, in the space of this document, by no means exhaustive it must be added that there are other opportunities (which the Council and London Transport have in hand) to improve the running and convenience of buses. Simpler and speedier fare collection, improved manning and better bus control systems seem to offer most. One-fifth of road passenger mileage in Greater London is by bus and the variety of effort to make the services effective and reliable reflects this important role.

Heavy lorries

6.10   Heavy lorries can be troublesome on city streets. This problem is worst in residential areas where these vehicles are out of scale with the general surroundings and can be dangerous. The Council, the London Boroughs Association and the Metropolitan Police have been looking into the problems of parking heavy vehicles in London. Some fourteen thousand lorries park overnight on London's streets but less than one thousand in lorry parks. There are about two-and-a-half thousand spaces in public off-street parks available at present. The boroughs with the worst problems include Brent, Camden, Hackney, Haringey, Islington, Lambeth, New-ham, Southwark, Tower Hamlets and Waltham Forest. The local lorry and the long-distance lorry pose separate problems, the first requiring small non-residential parks on a borough-by-borough basis and the second a regional network of properly constructed lorry parks with hostels for the drivers.

6.11   A pilot control scheme covers about one-and-a-half square miles in Tower Hamlets. It has been very successful—effective in removing the heavy vehicles and, with Police assistance using mobile wardens, easy to enforce. The authorities now intend to extend night-time lorry parking controls progressively over about sixty square miles in Inner London. The ban will act as a spur to lorry operators to make off-street arrangements for lorries commonly parked overnight on the street. Many local operators already accommodate their vehicles off the streets and this would only really amount to an extension of current practice. At the same time convenience and ease of enforcement would be increased if special lorry parks were provided in those areas. This may take some time and, to start with, in order to speed up control, lorry parking may have to be allowed in some non-residential streets.

6.12   Measures are therefore under way to deal with what is an annoying and disturbing practice. There seems no reason why any area which is troubled significantly by this problem should not be dealt with once the main effort on the worst trouble spots is made.

6.13   But it is not just parked goods vehicles that cause many people concern. While vehicles of a moderate size are common on our roads—and we all understand the important part they play in the servicing of the city—we now see very large vehicles indeed passing through some of our most prized areas and narrowest streets. Is it necessary that they should? In the long term the Council would certainly hold that most of them should use the ringways. In the meantime what could be done? At once we

26

have the now familiar environmental problem—the possibility that action in one area may transfer the problem to another—robbing Peter to pay Paul. But there are degrees of disturbance. Many would think that central London with its streets filled to capacity with pedestrians and buses, with its national monuments and the ever increasing numbers of visitors, is particularly unsuitable for this traffic. A small area in the City has been covered by a ban for some time. Would the right move now be to ban the largest vehicles from most of the rest of central London, certainly from areas such as Piccadilly Circus and Trafalgar Square?

6.14 But without the new ringways which the Council has proposed even this limited action will displace these lorries on to roads which are still not entirely suited to them. Enforcement might also be a problem, partly at least because of this. The same goes for other measures which might be thought of. For instance some might find it a boon if heavy vehicles were banned altogether from some quiet areas in the night hours.

7 The policies combined

The different combinations

7.1 What is the potential in the short term of the various options that have been described for dealing with traffic? In practice no single measure would be applied on its own. It is the effect of a combination of them which must be considered in different parts of London. It may be useful to discuss the situations in the Central Area on the one hand and in a suburban town centre on the other in order to see how different mixes of action might be taken and what they could achieve.

The Central Area

7.2 The centre of London is of national, indeed of international, importance. More than that it is the main driving force in London, the source of most of London's wealth and the key to our ability to achieve a high standard of life and environment. It can contain up to two million people at its busiest and has as much traffic on its streets as the whole of a city the size of Coventry—and that in spite of extensive Underground railways. It is in many ways quite natural for a metropolis to be busy and somewhat congested. Indeed, for all the critical comment about the situation on the streets, by comparison with the centres of other capital cities of comparable size London works freely and well and visitors often remark on this.

7.3 This is why some Londoners of long standing will say—" it was always busy and congested; it works well that way; don't kill the goose which lays the golden eggs by tampering with the situation ". However, others who see the large numbers of people on foot and are particularly conscious of our historic and architectural heritage would claim that central London needs very little private car traffic and few lorries. The questions therefore must certainly be asked—is there too much traffic on Central Area streets? Is all of it essential to the economy of the area? Even those who are concerned with traffic flow alone might now be tempted to say that there was too much; for although traffic on good days moves quite well the basic load on the system is so close to capacity that, if the weather is bad or there is an accident or some special event, big delays occur. If less traffic is desirable, how much less? Even a small reduction of up to ten per cent would make quite a difference particularly if it were in addition to the improvement which area traffic control will bring. Such reductions could give scope for further environmental protection as well as a far better flow of cars and buses and other traffic. A small change would have a limited effect on the key activities of London; but if such a moderate increase in restraint were to be made many people would want to see what the effects were before going too far.

7.4 If a move in this direction were thought desirable, where do we start from and how could it be achieved? The present general restraint of traffic to the Central Area is by congestion and parking control. Although congestion is thought by some to be a ' natural ' part of the scene, the uncertainties it brings about and the costs put on all users of roads make it something to be avoided if at all possible. For central London, parking control bears on little more than half the traffic, for about one-fifth of it is through-traffic and about a quarter is buses, taxis and goods vehicles.

7.5 The Council has always stressed that current parking targets were framed in the expectation of ringways at a fairly early date. Now, quite irrespective of the timing and content of a decision on the Greater London Development Plan, it is clear that only a part of this relief will accrue within the next ten years and a new look at the situation must be taken.

7.6 Present plans still envisage a slow increase in the total parking stock. Although on-street parking is already well down and could be reduced a little more, more public off-street parking is planned and, even at the new standards, car parks under new office blocks will still add significantly to the total number of parking places available. Even with extensive priorities for buses, which will give them some substantial advantage, the general outlook for improvement with only this level of action could be considered poor, particularly in environmental matters.

7.7   The first possibility which exists for doing more is to use the Council's powers to require licensing of public off-street car parks. These parks are just over a quarter of the total of non-residential parking and control in the way suggested earlier could reduce traffic at peaks by up to five per cent while increasing off-peak traffic by a lesser proportion. The difference between peak and off-peak traffic conditions in the Central Area and on its approaches would be narrowed and there would be a limited overall improvement in traffic conditions.

7.8   Such control of existing public off-street parking would not limit the slow growth in total daily traffic generation in the Central Area. This could be restricted by limiting the total amount of publicly-available parking through strict control over the development of additional car parks. In practice this would mean that no new public parking facilities should be allowed except in the few parts of the Central Area which are poorly served by existing public car parks or where existing private garages in office blocks are converted to public car parks. This policy, along with the completion of on-street control and the new office parking standards, would mean that non-residential parking supply in the Central Area would grow by about ten per cent over the next ten years. Residential parking supply would increase as well if present ideas persist. If however extra provision for residents who do not have their own garages were to be made in existing off-street car parks the net increase in total parking space might not be quite so great.

7.9   The other environmental measure which could be applied without excessive difficulty would be the banning of through heavy lorries from the Central Area. This would reduce the number of vehicles by only about two per cent but the size of these vehicles means that the benefits to traffic flow and the environment would be greater than this figure might suggest.

7.10   The first group of additional measures which could be considered for central London are therefore:

 i) control of public off-street car parks by licensing

 ii) restriction of the provision of additional public off-street car parks

 iii) banning of through heavy lorries.

An optional extra to this package could be a ring of bus lanes or other chokes to capacity on the radial approaches to the centre; this is a possibility which the Council is looking at. But while the core of the capital might benefit it could be at the price of serious congestion and consequential environmental damage in the areas around.

7.11   The collective effect of these three main measures is difficult to predict precisely. In the Central Area peak hour traffic would be reduced slightly and the growth in round-the-clock traffic checked. Traffic levels would be more uniform through the daytime and evening. The reduced numbers of heavy lorries would be a distinct advantage. If allied with the Council's committed programmes of bus lanes and area traffic control the prospect would be of a slight general improvement.

7.12   Outside the Central Area these policies would not have a marked effect. On the one hand the limitation of traffic travelling to central London, expecially in the peak hour, would be beneficial; on the other hand the displacement of some heavy vehicles would not. Round the boundaries of the control zone in particular there could be local problems. On balance this package of policies might give a small net benefit to conditions in inner London.

7.13   These ideas would leave the large number of private parking spaces under offices and at government buildings untouched. Further improvements to Central Area conditions by parking control, beyond those referred to above, would require some control over the use of these spaces. The only other feasible steps would be the elimination of temporary car parks and a further reduction in the number of meters.

7.14   Table 4 helps one to judge the significance of the various possibilities. The current parking stock is set out together with the implications for the future of present ideas. An alternative based on the measures set out in paragraph 7.10 is shown for comparison. The reduction in the total number of spaces will be noted; the much greater proportion under some form of public control would enable restraint to be more flexibly applied, for instance so as to reduce car commuting while still serving business during the day.

7.15   As discussed earlier in this paper, there is another range of possibilities for improving the balance between private and public car parking. At one end of the scale there is voluntary action and at the other a parking tax. What is likely to be in the best interests of London and acceptable to

Table 4   Effects of different parking policies in the Central Area

Type of parking	Number of spaces (thousands)		
	1971	1981(A)	1981(B)
Public (off-street)	30	43	39
Private (non-residential)	53	60	56
Meters (non-residential)	21	20	20
Free	2	—	—
Total (non-residential)	106	123	115
% of above subject to public control	28%	27%	51%
Residential (off-street)	17	26	26
Residential (on-street)	7	9	9
Total (residential)	24	35	35
Grand total	130	158	150

1981(A)—projection of present policies.
1981(B)—projection of policies as outlined in paragraphs 7.10.

its citizens? Whatever the verdict on this particular point, all parking measures suffer from the disadvantages mentioned earlier—they only act on part of the traffic. Increase of through traffic under the great pressures which will exist in this next ten years could erode some of the gain and indeed this appears to have been happening recently.

7.16 Should then the backing of parking policies by supplementary licensing be considered? This would be a radical innovation and there is a need for further study of feasibility. This could throw up difficulties. But it might give some certainty and flexibility of control of traffic in the centre, both of vehicles passing through and those whose journeys end there. Might it not be preferable to the pursuit of more difficult parking measures and continuing congestion?

7.17 So options do exist even for the familiar streets of central London. There are big issues involved but what matters most is what people want, how badly they want it and what sacrifices they are prepared to make.

7.18   Many of the major town centres in London are equivalent to the centres of large provincial towns in their size and the service they give to surrounding areas. Because they are set in a sea of hundreds of square miles of urban development their traffic problems are worse than those of a freestanding town and not so susceptible to simple remedy. The remedy which is most common in this country and throughout the world to the traffic problems of such centres is a by-pass or ring road. Regrettably most of the important London centres, placed as they are across the main lines of communication, do not have such aids yet although the Council's strategy for highway construction is directed at providing equivalent relief. This present absence of town centre by-passes itself ensures that the problems of this decade will be acute in many places although progress will be made in a few.

7.19   The deficiency cannot easily be made good, for alternative routes selected from the existing street pattern are always very likely to pass through the residential areas in which the centres are set. Neverthless any environmentally acceptable management schemes which can reduce the load on centres clearly have much to recommend them.

7.20   By comparison even with central London, the exposure of these centres and their surroundings to traffic growth is going to be severe and difficult to control. The extension of general measures like supplementary licensing to such areas in a relatively short time seems improbable even if it were to be accepted on a trial basis elsewhere. It seems then inescapable that, of the measures discussed already, emphasis must fall on physical management and parking controls, particularly at peak hours. Local measures may have a useful contribution to make to the general strategy of restraint as well as to local problems.

7.21   In most centres widespread on-street parking controls are made more difficult by the proximity of residential areas. Where traffic is almost certain to grow steadily, stricter limits on the provision of off-street parking may well have a key part to play. While the high proportion of through traffic in many centres will limit what can be achieved a local willingness to accept stringent controls might help achieve the first stages of environmental improvement.

7.22   As the general level of traffic will be difficult to control except by congestion it seems clear that buses must be kept flowing as freely as possible and bus lanes and other priorities are likely to suggest themselves on the approaches to the busiest areas. Some short term help to pedestrians might also be sought successfully, particularly by the more extensive linking of traffic signals. But any measures in and around the town centres which tend to increase congestion will also tend to put pressure on the surrounding localities. Experience shows that traffic will try to find easier routes through back streets. Street closures and other measures could give protection against this but only at the loss of some local freedom of movement. The more regulations that need to be made the greater will be the already formidable task of enforcement.

7.23   Can anything more be done which is going to help in the short term? Probably not. However the opportunity does exist to protect these local centres before their problems become acute everywhere. If the authorities had foreseen in the fifties the problems which office car parking now presents in central London they would almost certainly not have pursued their original policies. There is still time for the suburbs to prevent themselves being saddled too firmly with the problems of the centre in miniature.

8   Concerted action

8.1   The Council, the London boroughs, the Government, the Metropolitan and City Police and the people of London themselves are all in their different ways members of the partnership which can alone by concerted effort make the best of London in the next ten years. The Council has the duty to make it plain what could be done and to propose what should be done. This document is intended to set out the situation for discussion. The Council will also have the job of using its powers to further the action which lies within its province. Other agencies too have their individual parts to play.

8.2   The Government, which must have a general interest in local transportation policies and progress, is involved also as a promoter of national legislation and through its regulatory powers, if such issues as keeper liability for traffic offences, supplementary licensing or parking taxes are to be tackled. The further investigation of road pricing and the development of environmentally acceptable vehicles are also clearly matters in which the Government as a sponsor of urban research should be expected to take the lead for national as well as London purposes.

8.3   The London boroughs as local planning, parking and highway authorities have a key part in local traffic and environment matters. Co-operation between the Council and the boroughs is essential if there is to be consistency between local and London-wide action. Plans for local action, which give a firm basis for parking provision, environmental protection and improvement and are in line with the wider needs of buses and of traffic generally, would help a great deal if they could be available quickly. The tedious formalities of the conventional planning process are too slow to meet some of the most pressing needs.

8.4   The Police have a vital role in the enforcement of traffic regulations and in the operation of the road system. They desperately need the help of the reform of traffic law already referred to, for the present system is cumbersome both for the Police and the public. Unless this help can be given, the burden on them will grow and the sheer load of work will defeat any wish to extend the scope of the controls they enforce. But given action to change the law and a general public acceptance of any new measures which may be introduced, the continued co-operation and efficiency of the Police are good guarantees for the better management of London's roads.

8.5   People are the key element in the conflict between traffic and the environment. It is their changing ways of living which have led to the problems we now have. Quite small changes in habits would help ease the situation in the short term—for instance, travelling as much as possible outside busy hours and away from congested areas, choosing public rather than private transport whenever reasonably convenient, particularly when going to busy areas at peak hours, and consciously avoiding using local roads as through routes.

9   A review: the key issues

9.1   What does all this add up to? The outlook in its baldest terms is this—given continued growth in car ownership in the next decade and no new measures of control, London's roads will fill up still further with traffic. They are not likely to grind to a halt but the inconvenience and cost of using them will grow and traffic will invade more and more the quieter areas when it finds convenient routes through them. The general environment is bound to suffer.

9.2   Buses could be protected from this situation to some extent. But neither any feasible improvement in the services run by the transport undertakings nor any new construction possible on road or rail within the next ten years will change significantly the relation of traffic demand to the capacity of the city to absorb it.

9.3   Something could be done to change the present situation, particularly in the Central Area and the inmost parts of London, by making existing measures against cars and lorries tougher and introducing new ones. The main issues which have to be settled are:

  i) how far should parking controls be taken?

  ii) should supplementary licensing be given serious consideration?

  iii) how far should restrictions on the road go?—bus lanes, road closures, and the like.

9.4   But the situation is very tight. For every gain there will be a loss somewhere to someone. Sweeping changes in the centre brought in too quickly could upset the way London functions as a national and international centre and as a focal point for work and play for its citizens. Most of the gains and losses will be to different groups of people and the judgement the Council has to make is where to strike a balance. There may be better surroundings and easier travel for some; less convenient travel and changed surroundings for others. One element might mean small advantages for the public at large, another substantial loss for certain individuals. Improvement to the environment might attract some to live in London; restriction might encourage others to leave.

9.5   All the measures which might guarantee change in the situation involve interference with personal liberty and extension of already plentiful rules and regulations. The Council will want to see that any sacrifices people may opt for in this direction will lead not to pettifogging controls without effect but to real improvement.

9.6   Whatever the sacrifices however there is no promise here of a general transformation in the environment in the seventies. The gains will be small and hard won. The physical straitjacket in which London's past has left it prevents anything more. Is it not likely that if in the long term traffic restraint were to be allied with new roads and better public transport—all complementary to each other—the improvement in environment would be both widespread and welcome? This must be the hope and expectation for the eighties.

9.7 But the Council is seeking by the publication of this paper to stimulate discussion and debate of narrower issues. In the light of what people have to say it must quickly review and settle with other authorities involved what the programme of action for traffic restraint should be in the next ten years.

Designed by GLC Supplies Department (Printing and Graphic Design Division) and printed by C. F. Hodgson Ltd. K46451 /65513 /7.72   10,000.